Earth's Place in the Solar System

SCHOOL PUBLISHERS

Orlando Austin New York San Diego Toronto London

Visit *The Learning Site!*
www.harcourtschool.com

What Causes Earth's Seasons?

VOCABULARY

axis
rotation
revolution

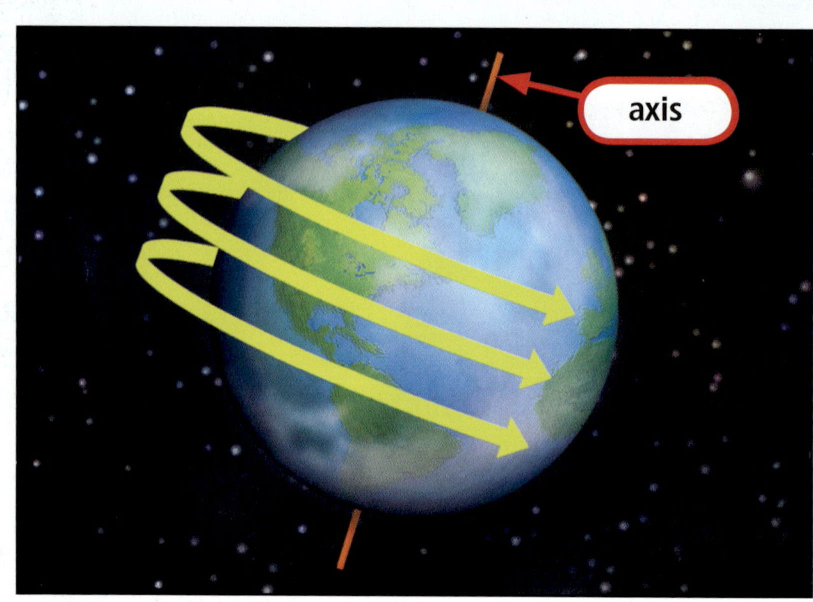

Earth's **axis** is an imaginary line that runs through the center of Earth.

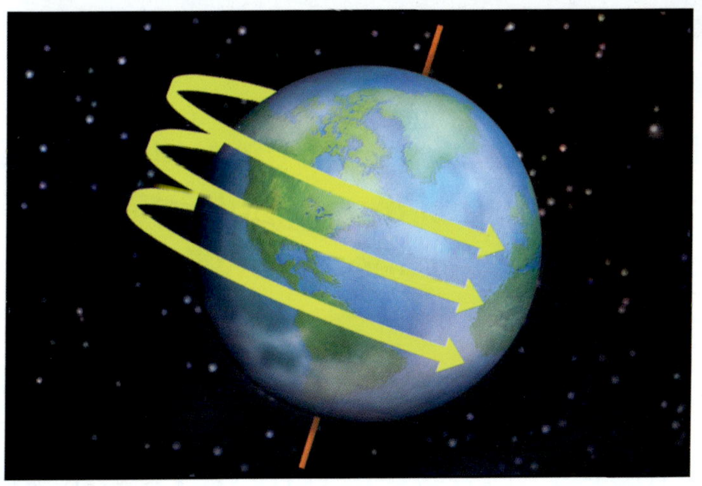

Rotation is the spinning of Earth on its axis. It takes Earth 24 hours to rotate once.

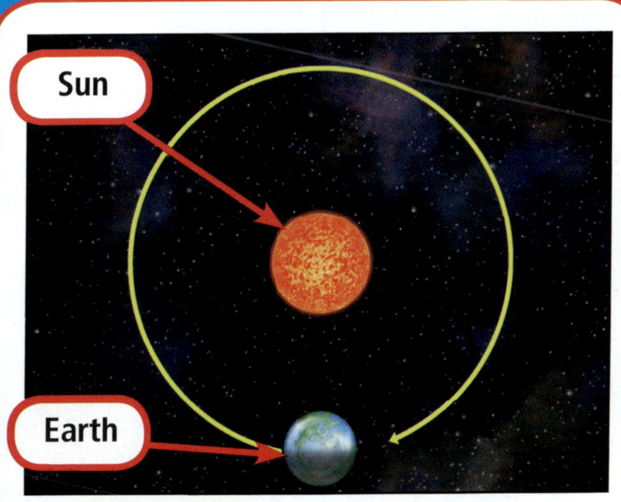

A **revolution** is the movement of Earth around the sun. It takes one year for Earth to make one revolution around the sun.

READING FOCUS SKILL

MAIN IDEA AND DETAILS

The **main idea** is what the text is mostly about. **Details** tell more about the **main idea**. Look for **details** about how Earth moves.

How Earth Moves

Earth moves in two different ways. One way is that it spins around like a top.

Earth spins on its axis. The **axis** is an imaginary line through Earth. It goes from North Pole to South Pole. Earth's axis is tilted. You cannot see the axis. The spinning of Earth on its axis is called **rotation**.

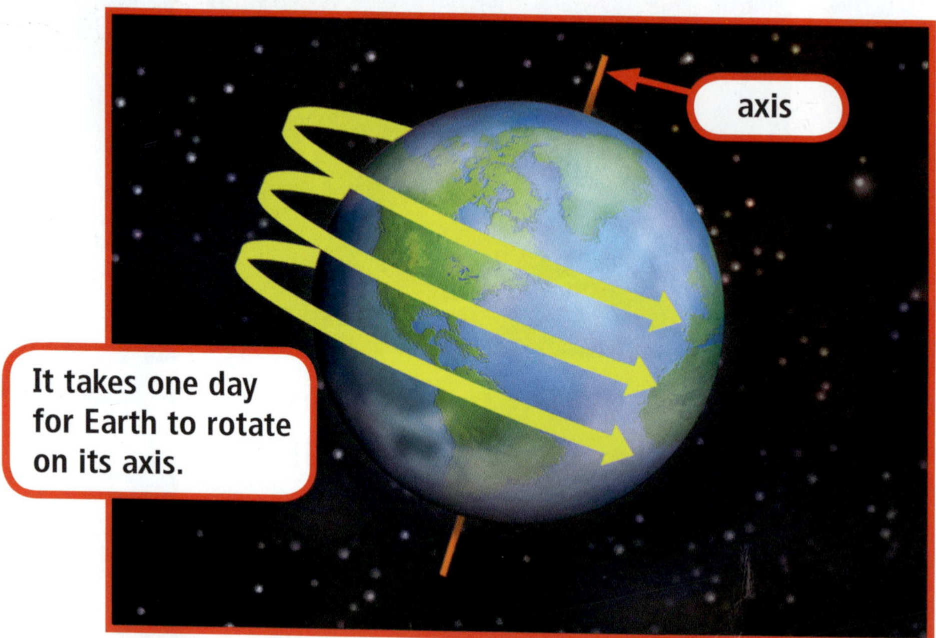

It takes one day for Earth to rotate on its axis.

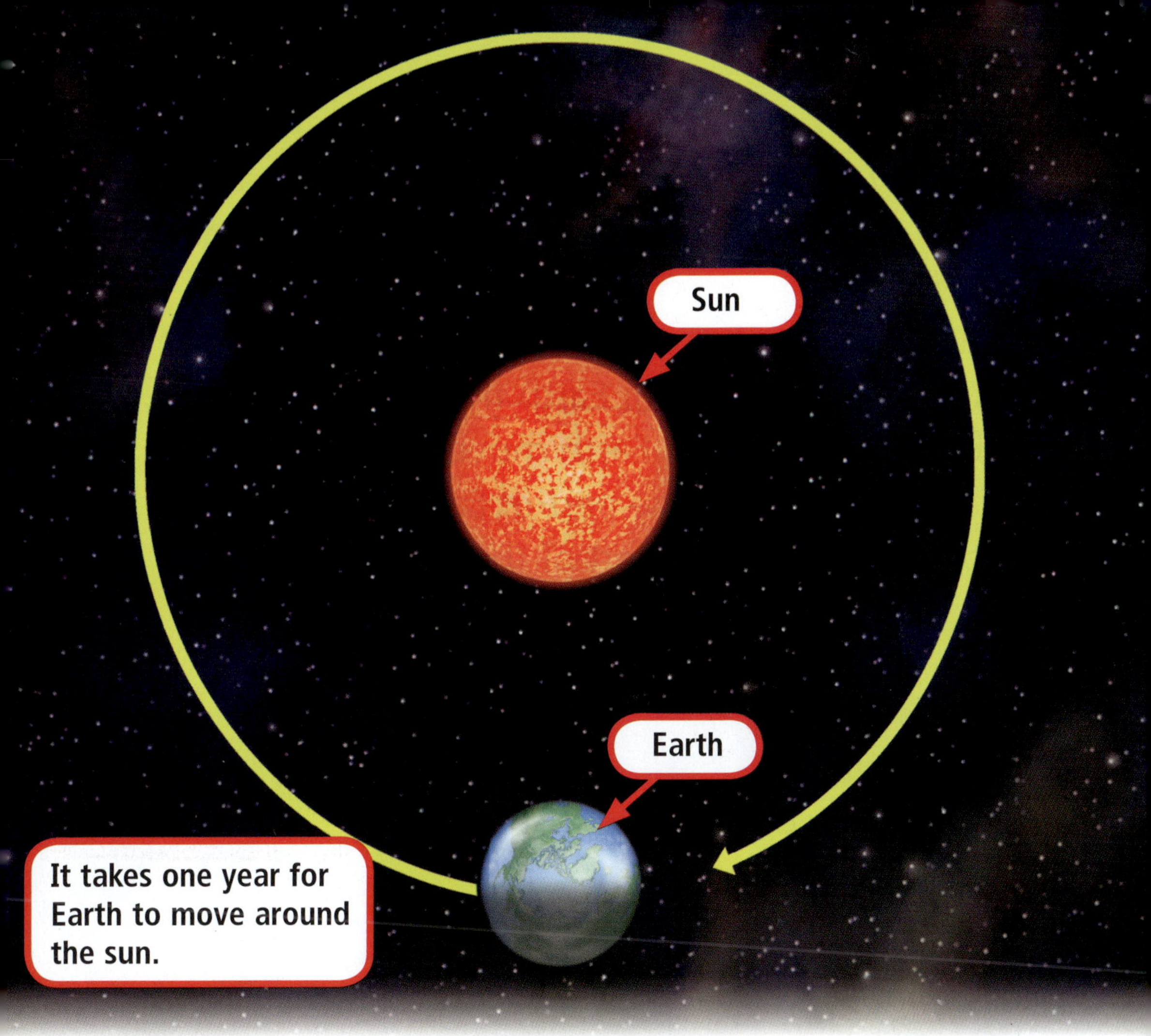

It takes one year for Earth to move around the sun.

Earth also moves around the sun. A **revolution** is one trip around the sun. Each revolution takes about 365 days. We use Earth's revolution to measure time. One revolution of Earth takes one year.

Tell the two ways that Earth moves.

Seasons

Many places have four seasons during one year. These seasons are winter, spring, summer, and fall.

Each season has a different temperature. That is because the part of Earth that is tilted toward the sun changes as Earth moves around the sun. This causes the same part of Earth to be tilted toward the sun for part of the year and away from the sun for part of the year.

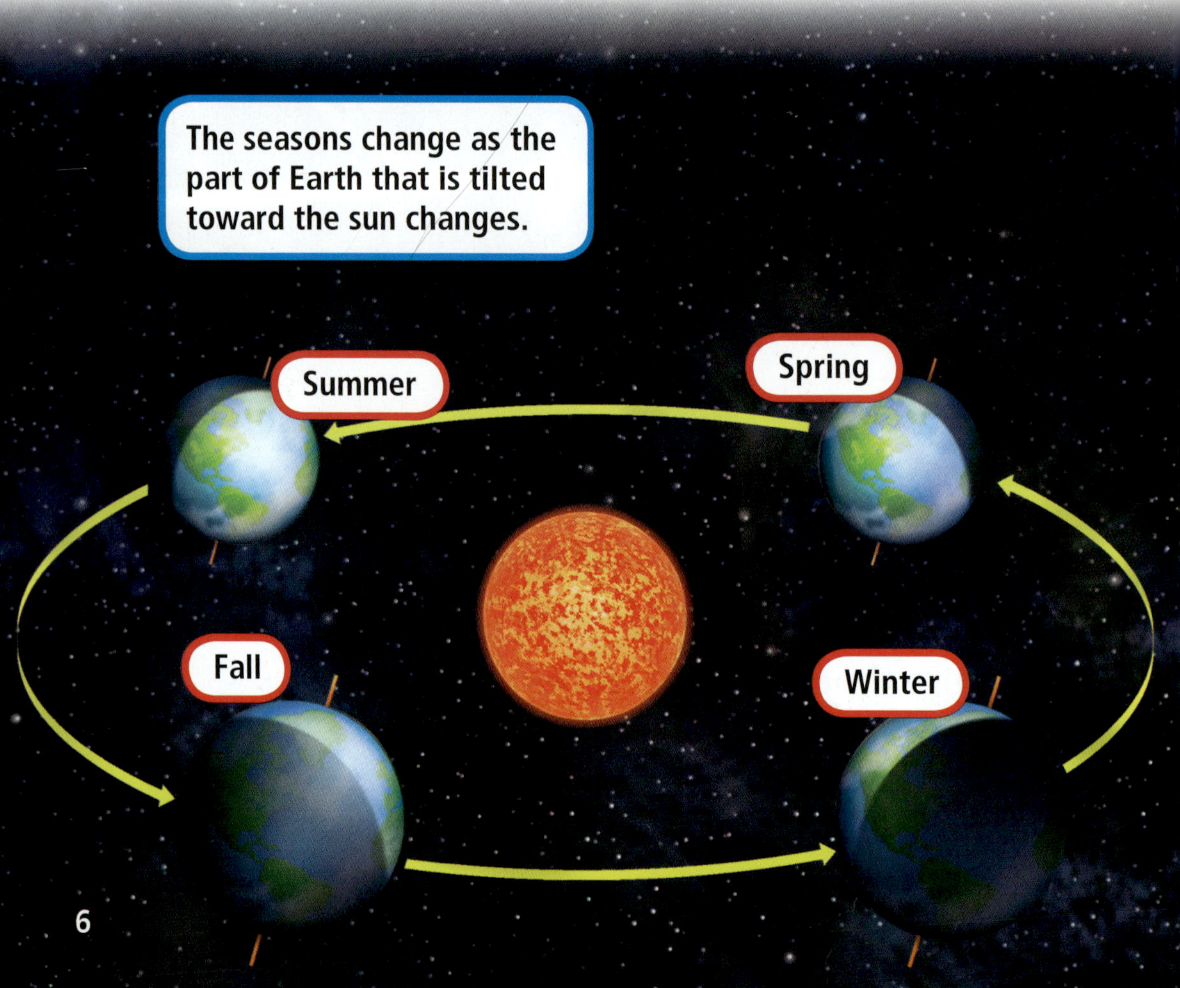

The seasons change as the part of Earth that is tilted toward the sun changes.

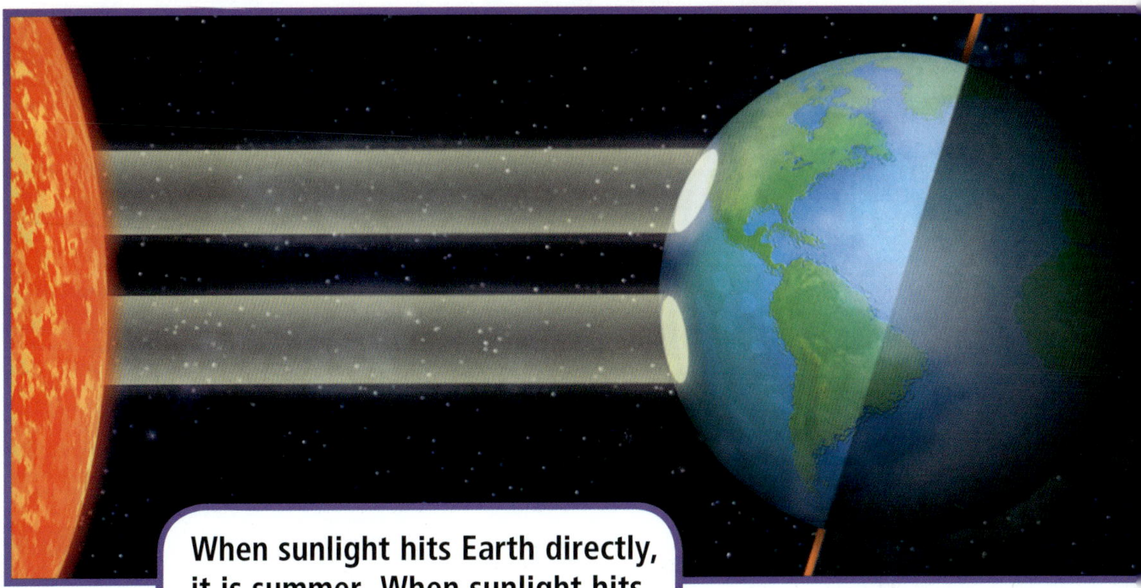

When sunlight hits Earth directly, it is summer. When sunlight hits at a slant, it is winter

When the part of Earth where we live is tilted toward the sun, we have summer. Sunlight hits that part of Earth directly. This makes it warmer.

When the part of Earth where we live is tilted away from the sun, we have winter. Sunlight hits that part of Earth at a slant. This makes it cooler.

Tell what causes the seasons to change.

Day and Night

Earth takes one year to move around the sun. But Earth takes only one day to rotate all the way around.

As Earth rotates, one side faces the sun. This side has day. The other side of Earth faces away from the sun. This side has night.

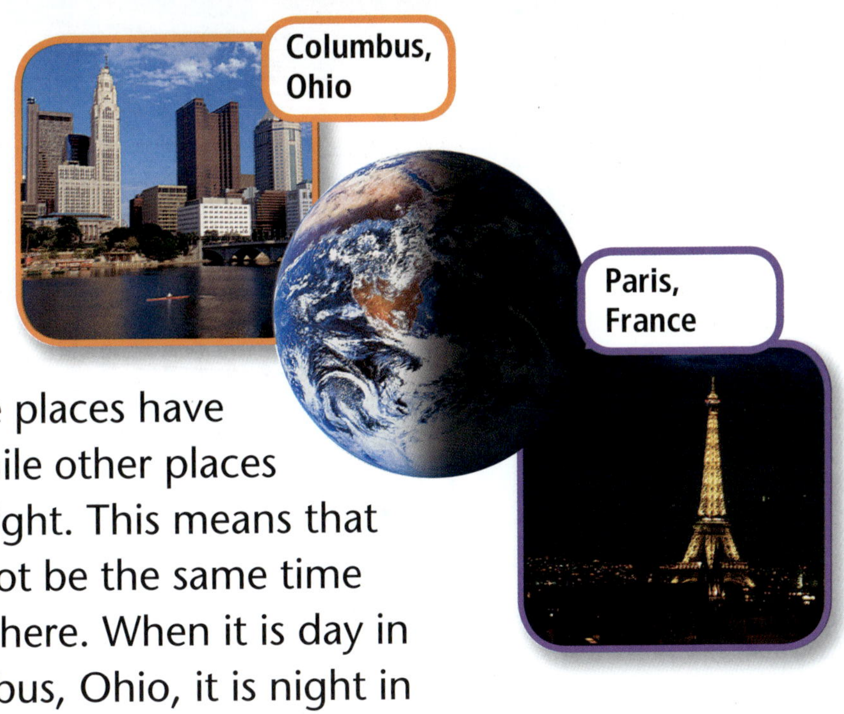

Some places have day while other places have night. This means that it cannot be the same time everywhere. When it is day in Columbus, Ohio, it is night in Paris, France.

What causes day and night?

Review

Focus Skill

Complete this main idea statement.

1. Earth moves in ______ different ways.

Complete these detail sentences.

2. Earth's ______ causes day and night.
3. It takes one _______ for Earth to revolve around the sun.
4. The ______ of Earth's axis causes the seasons.

How Do Earth and the Moon Interact?

VOCABULARY

moon phases
lunar cycle
lunar eclipse
solar eclipse

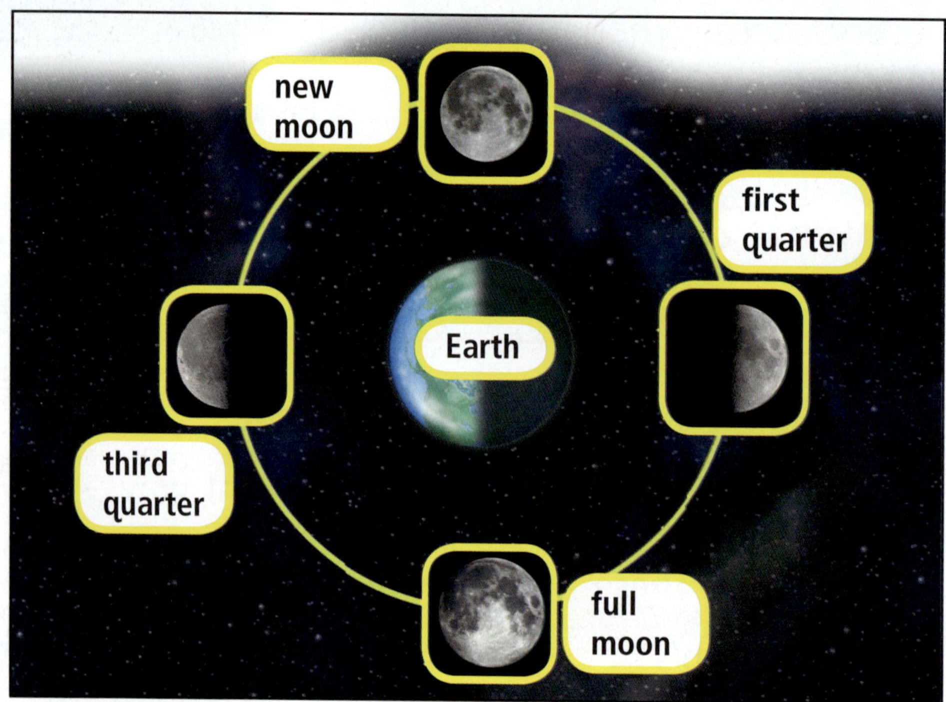

Moon phases are the different shapes that the moon seems to have. The moon's shapes follow a pattern that repeat about every 29 days. This pattern is called a **lunar cycle**.

A **lunar eclipse** happens when Earth blocks sunlight from reaching the moon. The moon gets darker when this happens.

A **solar eclipse** happens when the moon blocks sunlight from reaching Earth, and the moon's shadow falls on Earth.

READING FOCUS SKILL

SEQUENCE

A **sequence** is the order in which things happen.

Look for the **sequences** of the moon's phases and eclipses.

Phases of the Moon

The moon does not always look the same. From Earth, its shape seems to change. The different shapes we see are called **moon phases**.

Sun

In fact, the moon's shape does not change. It is the moon's movement around Earth that makes the moon's shape seem to change. The shape of the moon you see each night depends on how much of the moon's sunlit side faces Earth.

The moon's shape changes in the same order every 29 days. This pattern of changes is called the **lunar cycle**.

Tell what phase comes after the new moon phase.

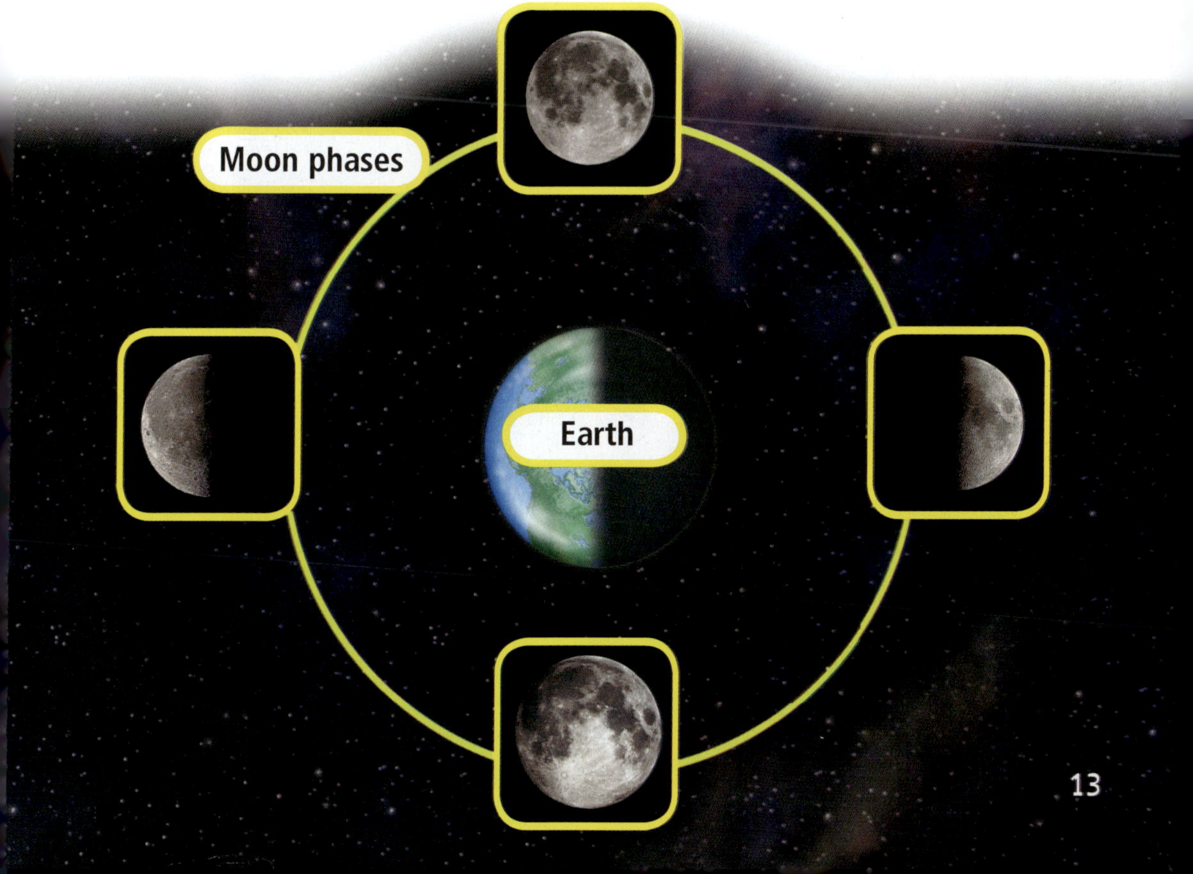

Moon phases

Earth

Eclipses of the Moon

Sometimes the moon moves into Earth's shadow. This causes a lunar eclipse. A **lunar eclipse** happens when Earth blocks sunlight from reaching the moon.

During a lunar eclipse, the moon first gets dark. This happens when the moon moves into Earth's shadow. Then the moon gets bright again. This happens when the moon moves out of Earth's shadow.

 **Tell how a lunar eclipse happens.**

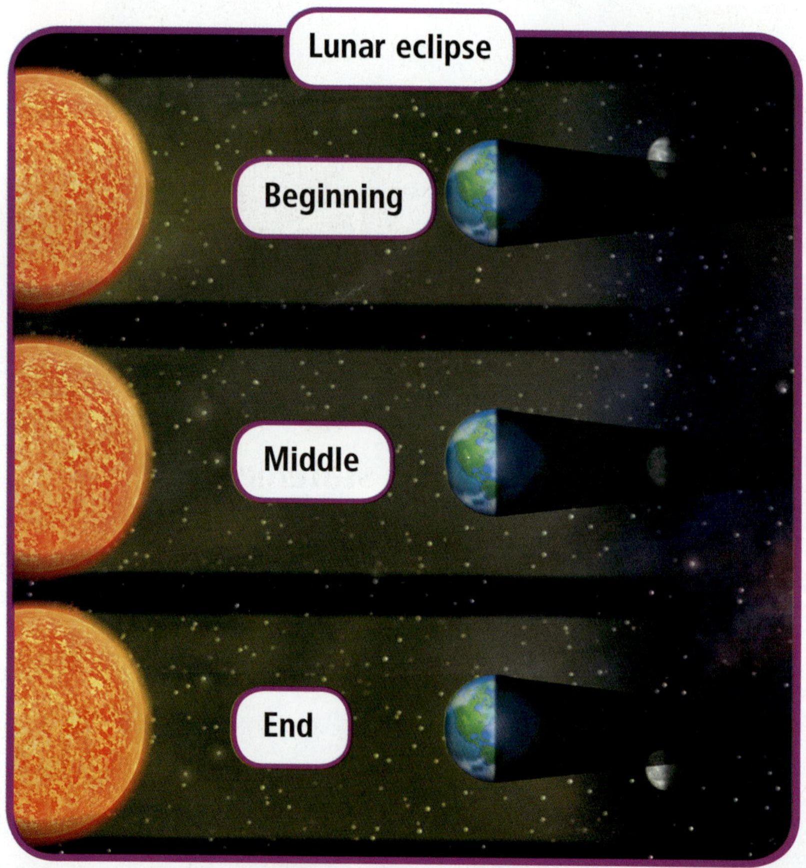

Eclipses of the Sun

The moon can block sunlight from reaching Earth. This causes a solar eclipse. A **solar eclipse** happens when the moon's shadow falls on Earth.

During a solar eclipse, the moon moves between the sun and Earth. This makes the sky get dark. As the moon moves on, daylight returns.

 **Tell how a solar eclipse happens.**

Solar eclipse

Review

Focus Skill **Complete these sequence statements.**

1. A ______ happens when Earth blocks sunlight from reaching the moon.
2. A ______ happens when the moon blocks sunlight from reaching Earth.

What Is the Solar System?

VOCABULARY

stars
planet
orbit
solar system
constellation

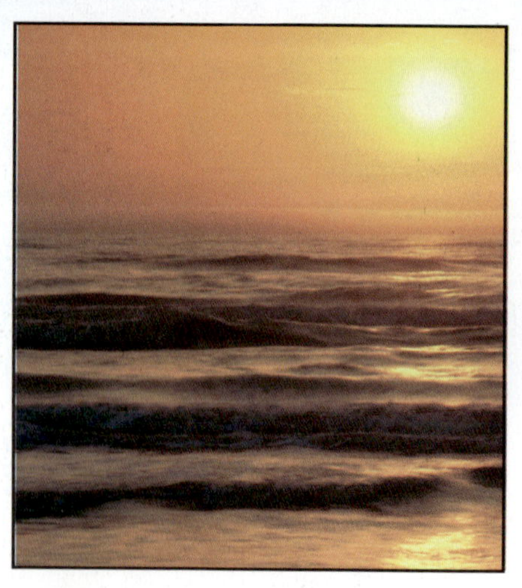

Stars are hot balls of gases that give off energy. Our sun is a star.

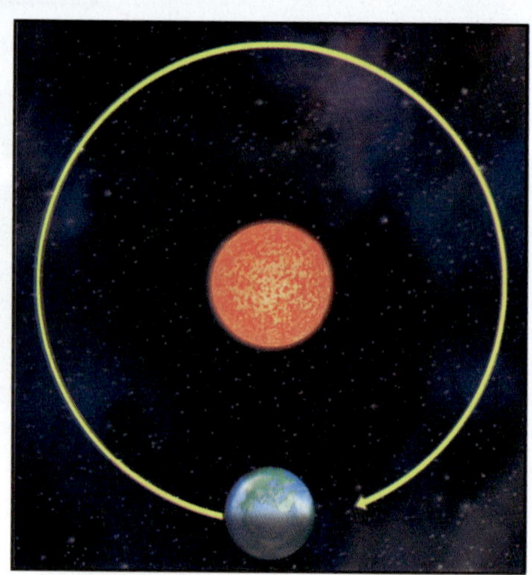

A **planet** is a large body of rock or gases in space. An **orbit** is a path that each planet travels around the sun.

The **solar system** is made up of the sun, the planets and their moons, and other small objects.

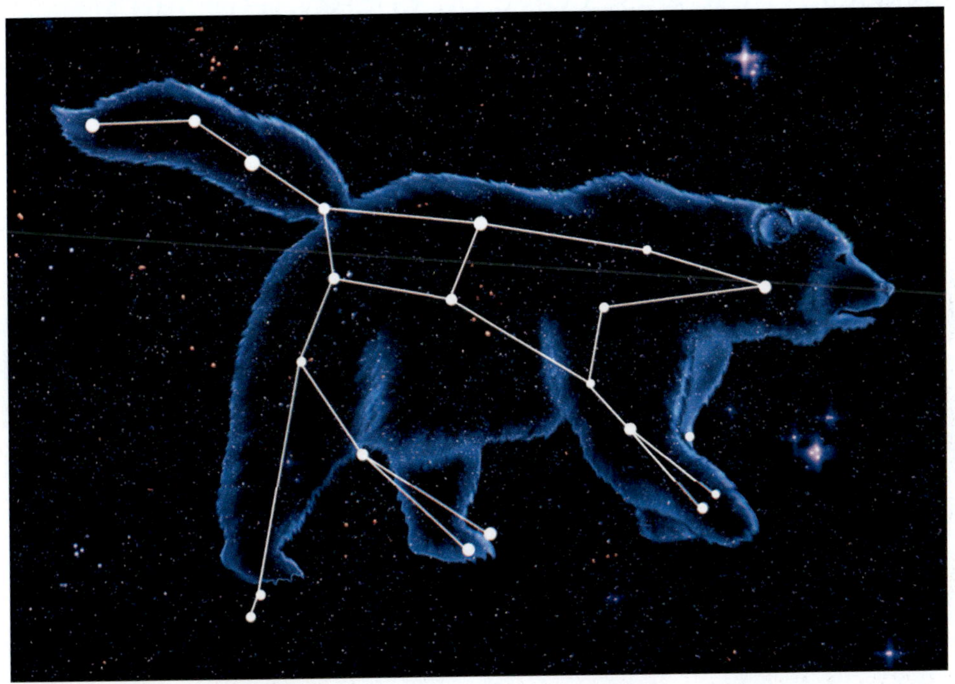

A **constellation** is a group of stars that forms a shape. The shape could be of a person, animal, or thing.

READING FOCUS SKILL

COMPARE AND CONTRAST

When you **compare and contrast**, you tell how things are alike and different.

Look for ways to **compare and contrast** planets in the solar system.

The Sun and the Solar System

The sun is a star. A **star** is a ball of burning gases. These gases give off energy. The sun's energy helps plants grow. It keeps Earth warm.

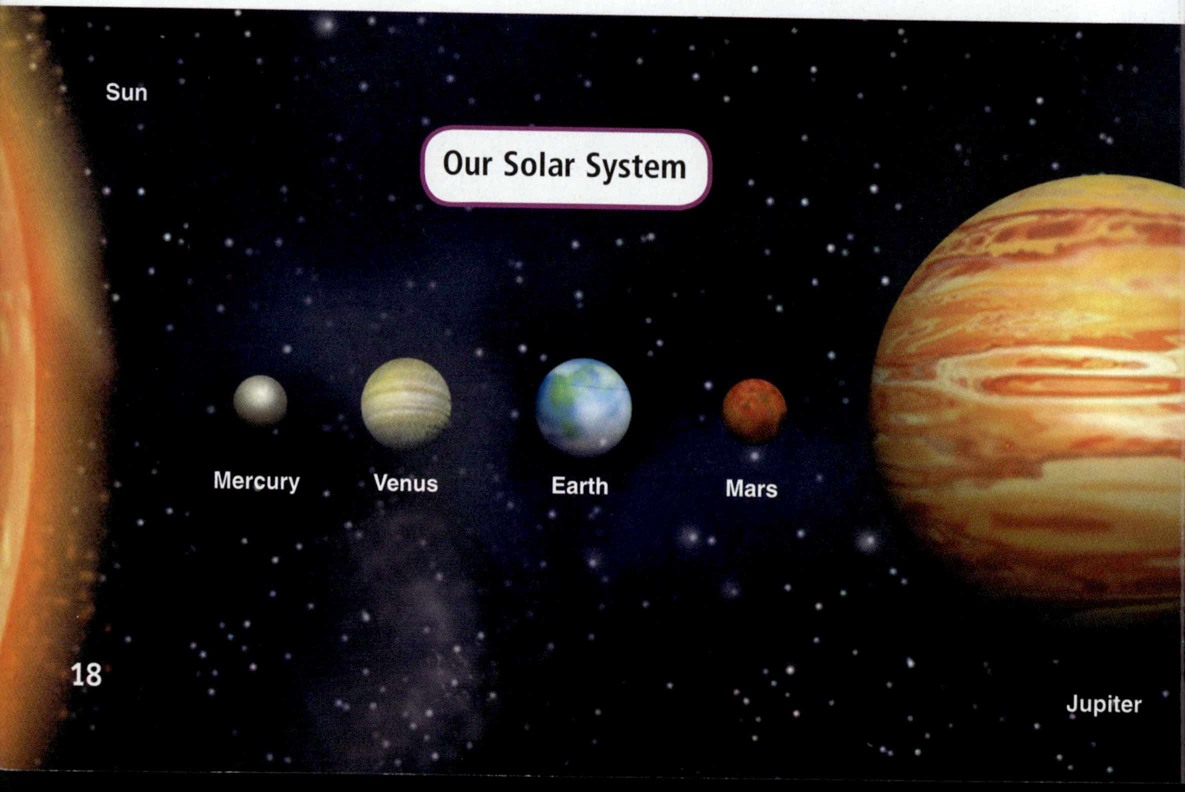

Our Solar System

Sun

Mercury

Venus

Earth

Mars

Jupiter

Earth moves around the sun. Eight other planets do this, too. A **planet** is a large body of rock or gases. Each planet travels in an **orbit**, or path, around the sun.

The **solar system** is made up of the sun, the planets, and other small things that orbit the sun.

Tell how all planets are alike.

Saturn

Uranus

Neptune

Pluto

The Inner Planets

The four planets that are closest to the sun are called the inner planets. These planets are Mercury, Venus, Earth, and Mars.

You can sometimes see Venus and Mars at night. They look like stars. But they do not twinkle.

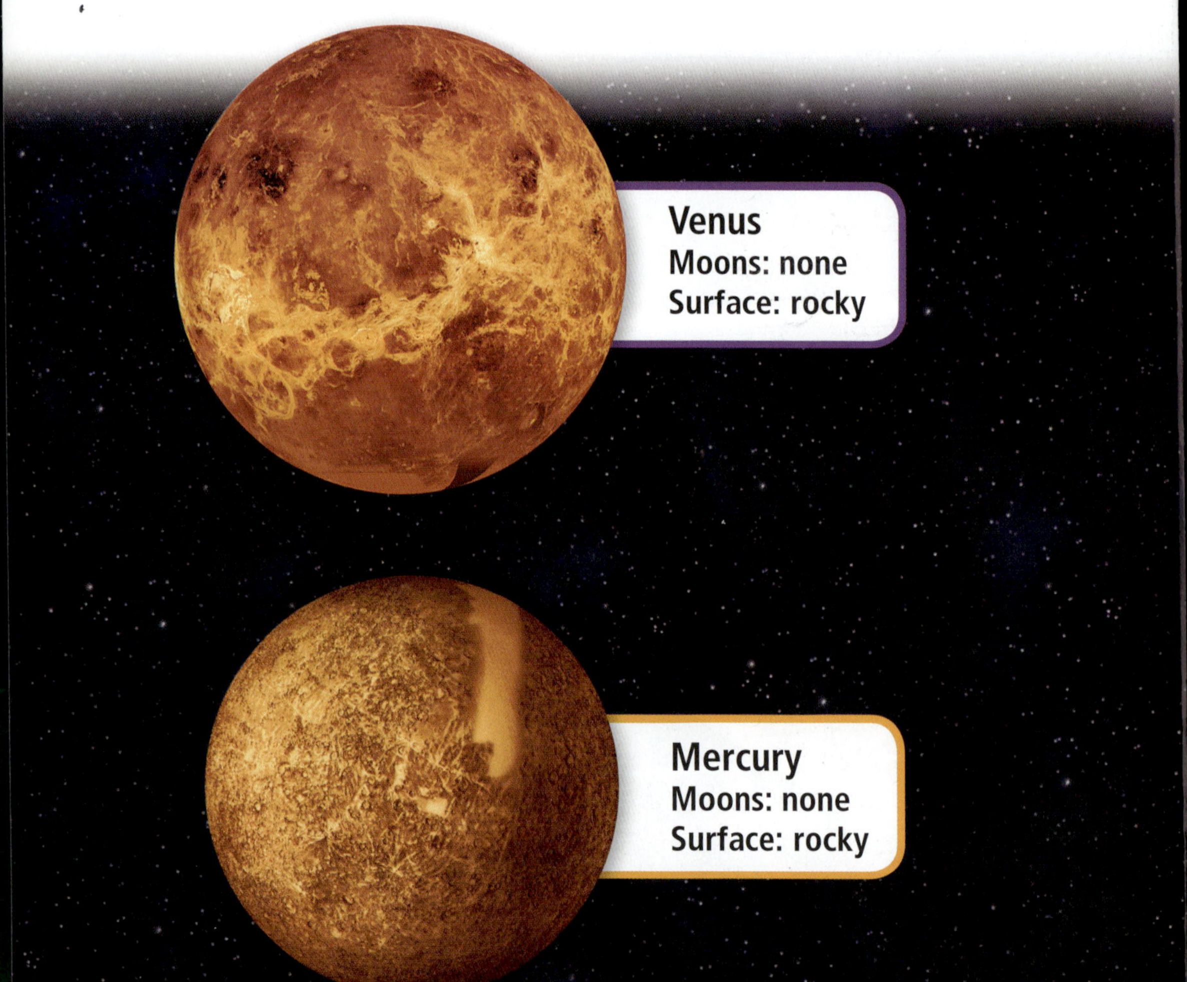

All the inner planets have rocky surfaces. They are smaller than most of the other planets. The inner planets are warmer than the other planets, too. This is because they are closer to the sun.

How are the inner planets different from the other planets?

Earth
Moons: 1
Surface: mostly water, with some land

Mars
Moons: 2
Surface: rocky, with red dust and no water

The Outer Planets

There are five outer planets. They are farthest from the sun. The outer planets are Jupiter, Saturn, Uranus, Neptune, and Pluto.

The outer planets are made mostly of frozen gases. Most are large and have many moons.

Tell how most outer planets are alike.

Uranus
Moons: at least 26
Surface: frozen gases

Jupiter
Moons: at least 61
Surface: no solid surface

Neptune
Moons: at least 13
Surface: frozen gases

Saturn
Moons: at least 33
Surface: frozen gases

Pluto
Moons: 1
Surface: frozen gases

Patterns of Stars

How do people tell one star from another? One way is to use constellations. A **constellation** is a group of stars. The stars appear to make a picture in the sky.

Compare how people group stars with how they group planets.

▼ Big Bear constellation

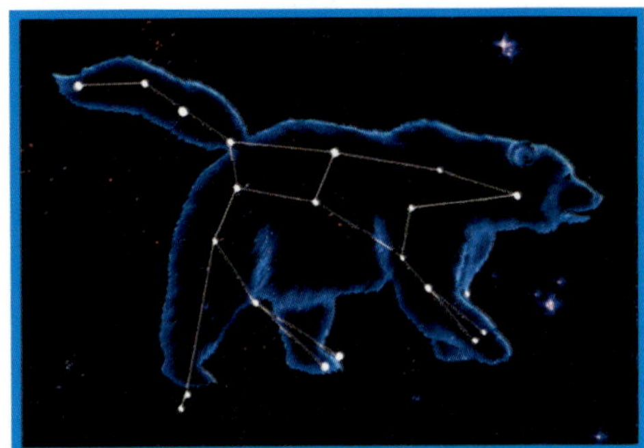

Review

Complete these compare and contrast statements.

1. All of the planets in the solar system ______ the sun.
2. The four inner planets are smaller than most of the ______.
3. Most of the outer planets have many more ______ than the inner planets.
4. All of the outer planets are farther from the ______ than the inner planets.

GLOSSARY

axis (AK•sis) an imaginary line through the center of the Earth

constellation (kahn•stuh•LAY•shuhn) a group of stars that appear to form the shape of an animal, a person, or an object

lunar cycle (LOON•er CY•kuhl) the pattern of phases of the moon

lunar eclipse (LOON•er I•KLIPS) an event in which Earth blocks sunlight from reaching the moon

moon phases (MOON FAYZ•uhz) the different shapes that the moon seems to have in the sky when the moon is observed from Earth

orbit (AWR•bit) the path that a planet takes as it revolves around the sun

planet (PLAN•it) a large body of rock or gas in space

revolution (rev•uh•LOO•shuhn) the movement of Earth around the sun

rotation (roh•TAY•shuhn) the spinning of Earth on its axis

solar eclipse (SOH•ler ih•KLIPS) an event in which the moon blocks sunlight from reaching Earth and the moon's shadow falls on Earth

solar system (SOH•ler SIS•tuhm) the sun, the planets and their moons, and the small objects that orbit the sun

stars (STARZ) hot balls of glowing gases that give off energy